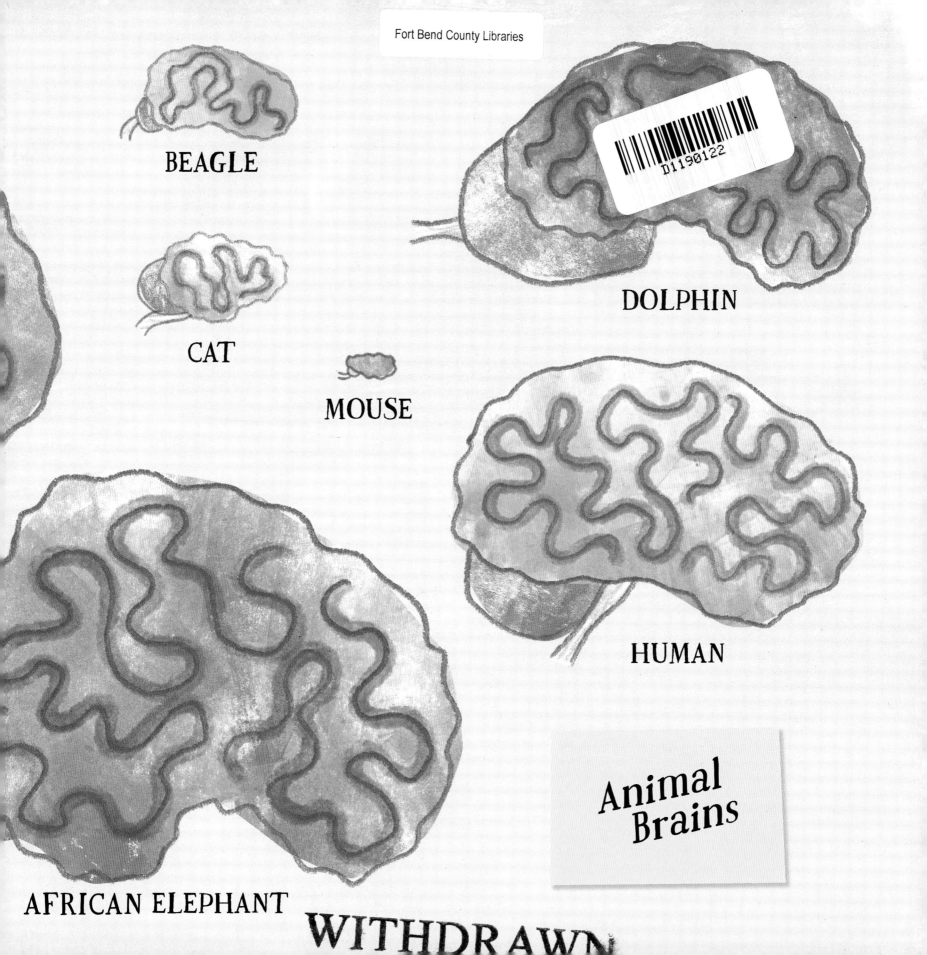

BEAGLE

DOLPHIN

CAT

MOUSE

HUMAN

AFRICAN ELEPHANT

Animal
Brains

For my dad, who loves the human brain
as much as a zombie would — S. M.

For Gary, who always
feeds my brain with
wisdom, humor, and
kindness — M. R.

Henry Holt and Company, Publishers since 1866
Henry Holt® is a registered trademark of
Macmillan Publishing Group, LLC.
120 Broadway, New York, NY 10271 · mackids.com

Text copyright © 2021 by Stacy McAnulty
Illustrations copyright © 2021 by Matthew Rivera

ISBN 978-1-250-30404-9
Library of Congress Control Number: 2020919359

Our books may be purchased in bulk for promotional,
educational, or business use. Please contact your local bookseller or
the Macmillan Corporate and Premium Sales Department at
(800) 221-7945 ext. 5442 or by email at
MacmillanSpecialMarkets@macmillan.com.
First Edition, 2021
Printed in China by Hung Hing Off-set Printing Co. Ltd.,
Heshan City, Guangdong Province

1 3 5 7 9 10 8 6 4 2

BRAINS!

Not Just a

ZOMBIE SNACK

Stacy McAnulty

Illustrated by
Matthew Rivera

GODWINBOOKS

Henry Holt and Company

New York

Don't worry.

I'm not going to eat your brain. I gave that up hours ago.

Human brains used to be my favorite food.
(About 75 percent water, 100 percent delicious.)

The brain is not a meal,
I keep telling myself.

And it is not a muscle.

Your heart is a muscle. And you have muscles in your arms and legs, fingers and toes, stomach and back, and even your face. You have more than 600 muscles.

But only one scrumptious brain.

Your brain is
the boss,

the coach,

the captain,

the command center,

the main computer—but
tastier than a computer.

OPEN!
ALL BRAINS
WELCOME!

Here's a list of things your body can do without your brain.

And here's a list of things that are yummier than your brain.

Yep, you're pretty useless without your brain.
And much less appetizing.

If someone nibbled on your brain, you wouldn't even know it. Your brain can't feel pain. It can't feel anything!

But if a vampire nibbled on your thumb, you'd know that.

The nerves in your thumb would send a message to your brain,

The brain receives messages from your

FIVE SENSES.

Sound: Your ears can hear a zombie sneaking up behind you. Unfortunately.

Touch: Your skin is your biggest organ, and it feels things.

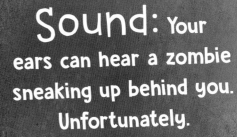

Sight: Your eyes can see a zombie approaching from the front. Unfortunately.

Smell: Your nose isn't just for picking.

Taste: Your brain is tasty. Er, I mean, you taste with your tongue.

I need a snack. I want some
brai . . . bran flakes.

Unless you know a grown-up with
a bit of brain to spare.
An adult brain is about three
pounds of yumminess.

A sperm whale's brain is about 17 pounds, but that's actually small compared to the size of a sperm whale.

Plus, I can't swim.

SPERM WHALE
17 POUNDS

AFRICAN ELEPHANT
11 POUNDS

BEAGLE
.16 POUNDS

MOUSE
.0009 POUNDS

OWL
.004 POUNDS

LION
.5 POUNDS

HUMAN
3 POUNDS

VIPER
.0002 POUNDS

So what makes human brains
yummier than whale brains?

Let's take a closer lick—I mean LOOK . . .
a closer look at the human brain.

NEURONS! They do the work, and you have about 86 billion of them.

And **GLIAL CELLS.** They feed and support the neurons. You have about 86 billion of them, too!

You've had most of these
neurons since you were born!

(They were yummy then.
They're yummy now.)

Each neuron can send
thousands of messages to other
neurons, creating pathways.

When you learn
something new, like
how to . . .

add numbers,

Your brain started forming when you were about half an inch long.
Do you plan on keeping your big, juicy, delectable brain your whole life?

Grown-Up Human

Brain: About 3 pounds
About 86 billion neurons

75-Year-Old Human

Brain: 10 percent smaller than max size (so 2.7 pounds-ish)
About 86 billion neurons

Newborn Baby

Brain: Less than 1 pound
About 86 billion neurons

One-Year-Old Human

Brain: About 2.2 pounds
About 86 billion neurons

I've got an idea.

Perhaps you'll offer me a teeny-tiny taste of your brain
if I promise not to eat any of the important parts.

I won't eat your brain stem.

You need this to breathe, and to hiccup, and to sneeze, and to keep your temperature at 98.6 degrees Fahrenheit, and to make sure your heart is beating.

All the stuff you don't think about.

APPETIZERS

BRAIN STEM

100% organic

Plus I find it a bit chewy.

And I won't nibble on your cerebellum, either. It's needed for balance and to coordinate your muscles.

Like, say, if you want to run away from a zombie later.

You'll want your cerebellum.

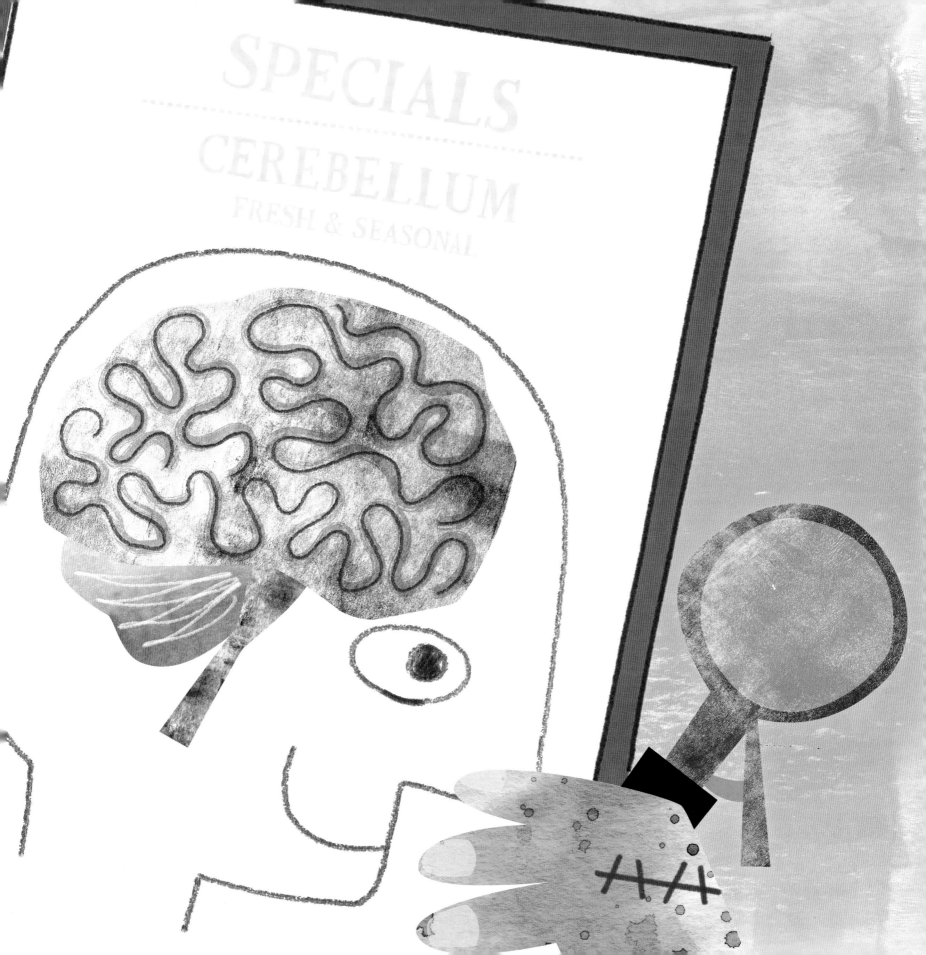

I like the cerebrum—the gray stuff.

It's the biggest part and comes in two halves. Left and right.
Things divided in halves are meant to be shared, don't ya think?
The cerebrum is filled with good stuff like seeing, hearing, emotions,
problem-solving, imagination, and memory.

Sweet, sweet memories.

And the cerebrum is mouthwateringly wrinkled.

CEREBRUM

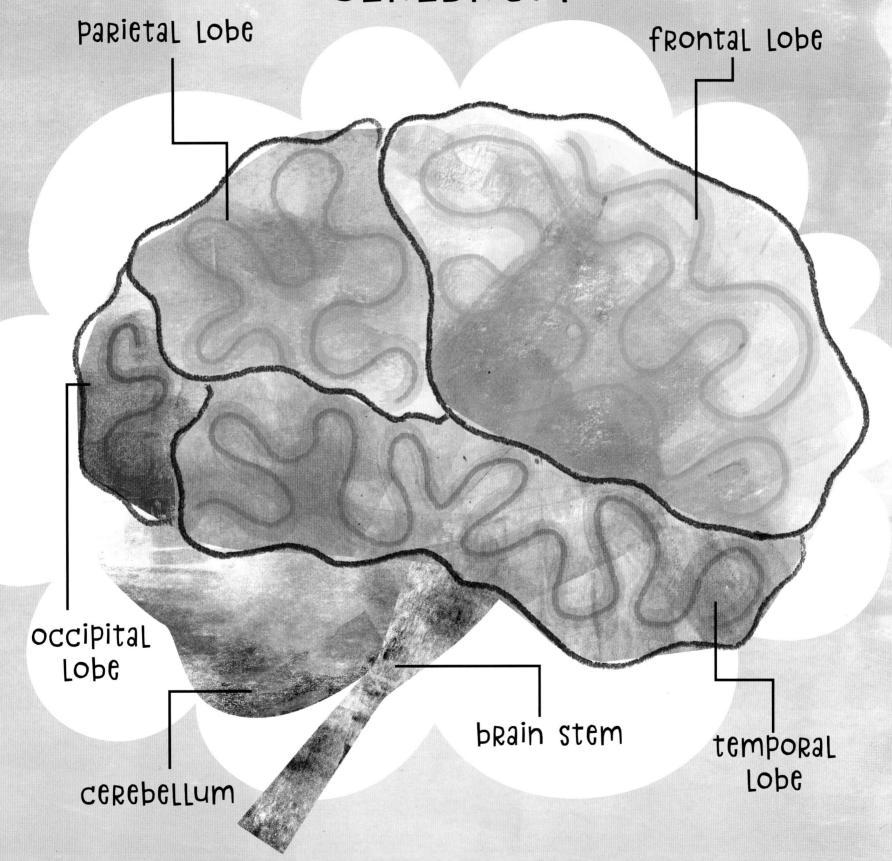

parietal Lobe

frontal Lobe

occipital Lobe

temporal Lobe

cerebellum

brain stem

Less-smart animals have smoother brains.

Mouse and bunny brains have hardly any wrinkles
and hardly any flavor. Blech!

I'd rather eat brussels sprouts
than rodent brains.

MOUSE

CAT

RACCOON

NOW SERVING
49
TAKE A NUMBER
3.9
20.00
Perfect!
HUMAN
BEAR
DOLPHIN
CAULIFLOWER

By eating your brain, I would be doing you a favor.

Your brain uses a lot of energy—enough to power a light bulb.

So whaddya say?

Please, please, please
share just a bite.

Or a spoonful.

Or a lick.

Can I just *smell* your brain?

FINE!

Keep your brain. Keep all of it.

I just hope you do something great with it

(like write a book about zombies).

Dear Reader with a Brain,

Do not give any of your brain to the zombie. Sure, she's hungry and the nicest zombie you'll ever meet, but you need your brain. You may have heard that you only use 10 percent of your brain, but that's not true. You use all of it, just not all at once.

The study of the brain is called *neuroscience*, and it's a fairly new and exciting field. Most of what humans know about the brain has been learned in the past decade. New computer technology is allowing scientists to look closer and closer at the delicious organ between our ears.

–Author and Not a Zombie
Stacy McAnulty

BRAIN FOOD
A Zombie's Guide to the Kitchen

Brain Facts:

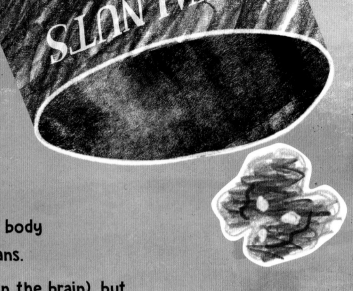

- A human brain is only about 2–3 percent of an adult's body weight, but it uses almost 20 percent of the body's energy.

- The tree shrew has a brain that is 10 percent of its body weight. But tree shrews are not smarter than humans.

- Humans are born with most of their neurons (cells in the brain), but they are capable of growing new ones. This is called *neurogenesis*.

- After he died in 1955, Albert Einstein's brain was stolen by Dr. Thomas Harvey and cut into 240 blocks to be studied. Scientists are still debating if Einstein's brain is significantly different from an average adult's.

- Jellyfish don't have brains. They have nerve fibers.

- The cerebrum in humans is bigger than in other animals. Maybe that's why we are better at video games.

- Zombies aren't real, but legend tells us that their favorite food is brains. Yuck.

Sources:

American Association of Neurological Surgeons. "Anatomy of the Brain." Accessed March 31, 2020. aans.org/en/Patients/Neurosurgical-Conditions-and-Treatments/Anatomy-of-the-Brain.

BrainFacts.org. *Brain Facts: A Primer on the Brain and Nervous System*. Washington, DC: Society for Neuroscience, 2018.

Carter, Rita. *The Human Brain Book*. New York: DK Publishing, 2019.

Eagleman, David. *The Brain: The Story of You*. Edinburgh: Canongate Books, 2016.

Encyclopaedia Britannica. "Brain." Accessed March 5, 2019. britannica.com/science/brain.

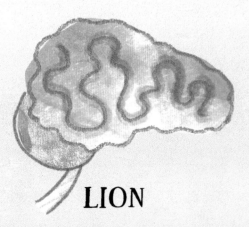

LION

RAT

VIPER

RACCOON

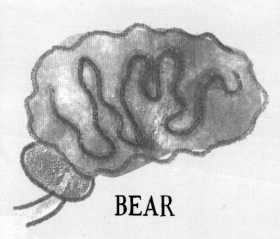

BEAR

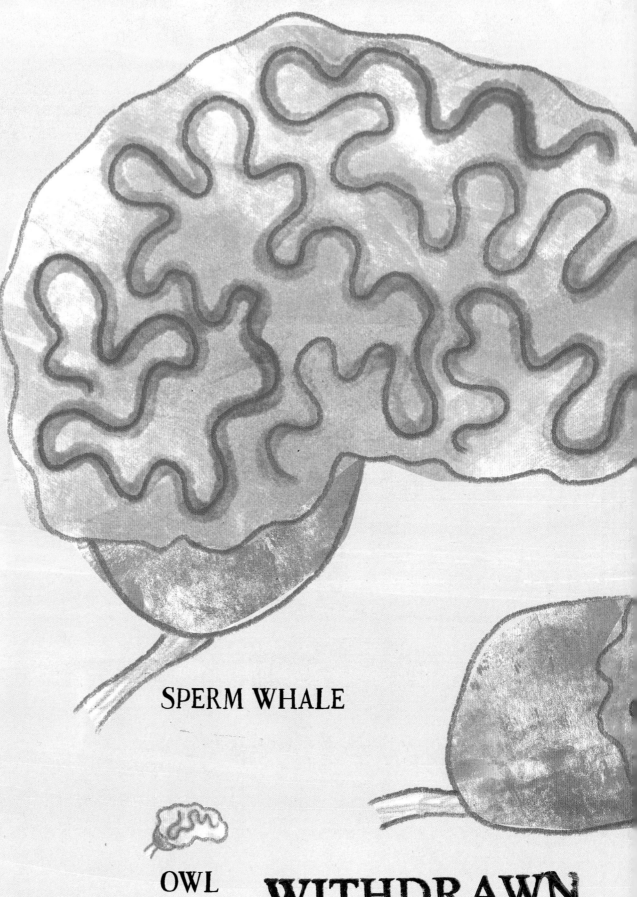

SPERM WHALE

OWL

WITHDRAWN